Edward Charles Spitzka

The Legal Disabilities of Natural Children Justified

Biologically and Historically

Edward Charles Spitzka

The Legal Disabilities of Natural Children Justified Biologically and Historically

ISBN/EAN: 9783337816759

Printed in Europe, USA, Canada, Australia, Japan

Cover: Foto ©berggeist007 / pixelio.de

More available books at **www.hansebooks.com**

THE

ALIENIST AND NEUROLOGIST.

VOL. XX.   ST. LOUIS, OCTOBER, 1899.   No. 4.

THE LEGAL DISABILITIES OF NATURAL CHILDREN JUSTIFIED BIOLOGIC-ALLY AND HISTORICALLY.*

By E. C. SPITZKA, M. D., New York.

(Continued from April, 1899.)

The "Grande Monarque" and his successor set examples which the petty princes of neighboring states were as quick to follow as had been Monseignieur and Marquis at home. Most of their Liliputian domains boasted an imitation Versailles and *parc aux cerfs*. This had been an unfaithful copy, had it not been stocked with game "of every shade." Foci of bastardization thus dotted the land so closely as to approach the confluent, and new sources of revenue must needs be, and unprecedented ones were discovered, to support these institutions of patriarchal, if not paternal government. The stipends granted by England in consideration of the loaned cargoes of food for the powder of "76", consigned as Hessian, Brunswick, Anspach and other mercenaries, went largely to support the bantlings of more than one duke and of two electors; one of these had no less than a hundred to feed! It was within the means of these princes to copy the architecture and garden, the costumes, menüs and etiquette of their Bourbon ideals; but, granting it had been their will, it was not always in their power to prevent the copies from degenerating in faithful imitation of their models, and, in one instance at least, from reproducing the most shameful features of the original. If the *Dragonades* of the Cevennes, the broken heart of Vauban, the humiliation of Racine, the exile of DuQuesne and others of the best brain and blood of France were due to the bigoted adventuress who wielded

*All Tables to be found at end of article.

despotic power over its king, through having been the governess to his natural children, no less was the startling reproduction of mediaeval despotic *camarillas* in the Bavaria of last century, due to an intrigant fanatic;[51] whose power had its source in the studied sycophant interest he affected for Charles Theodore's natural children.

The objects of a paternal interest so intense, so general and often so strangely manifested, are not the beings, whose helplessness might justify the exaggerated solicitude shown them. They commonly possess more than the average share of the faculty which ensures care of self. Be it through moral or wilful deafness to those behests of Law, which restrict within reasonably fair bounds the civilized strugglers in the contest for existence, the bastard, keenly sensitive to emergencies, intensely appreciative of momentary opportunity and above all, robustly selfish, inclines to depart from the conventional and slow into more opportune and shorter paths. Competition he may overcome by a radical and effective, howbeit irregular method, removing it altogether by destroying the competitor; one perhaps like Edgar in Lear:

"* * * * * a brother noble"
"Whose nature is so far from doing harms"
"That he suspects none; on whose foolish honesty"
"My practices ride easy! I see the business"
"Let me, if not by birth, have lands by wit"
"All with me's meet, that I can fashion fit."

It is rapacious ambition, leading to spurious claims, that accounts for the enormous proportion in eighty-four homicides among four hundred bastards of thirty-two (including sixteen fratricides) who had murdered relatives of the first degrees of consanguinity; that is thirty-eight per cent (Table I). Whether it were one *nothus* who, like Gideon's, found seventy(!)legitimate brethren in his way, or seventeen bastards, who wistfully regarded Artaxerxes throne out of which they were kept by the single legitimate heir, the *filius nullius* was equal to the occasion; witness the sixty-nine victims of fratricide Abimelech and the murder of Xerxes the Second, inaugural of a chain of assassinations and executions of brother by brother continued in the death of his murderer, bastard Sogdianus, at the instance of Darius nothus, and the killing later on of brother nothus Arsites. Need I more than refer to the historically famous victims of fratricide bastards from Remus of Legend and Amnon of Scripture, to Roman Geta, Macedonian Demetrius, Pedro of Castile and Juan of Gandia?

It is true that like tragedies frequently occurred in ruling families, committed by princes of legitimate birth, as well as by their illegitimate half brethren. But while, in the former case, the crime has been sometimes punished by the unfortunate father common to victim and assassin, yet although the deed is proportionately much more frequently a bastard's, I fail to find one instance where paternal partiality had become sufficiently neutralized by grief for the slain and resentment at the offender to ensure the latter's adequate punishment. Many princes, who had not been guilty of so grave or indeed of any serious misdeeds, have been executed or assassinated by command of suspicious, deluded or cruel royal fathers.[52] But only one instance can I find, where a bastard son was the victim: It is that of a paramour of his father's second wife, the Parisina of Byron. So bitter was paternal remorse however, that it sought relief in expiatory sacrifices — as usual with princes — of others; for Nicolo of Este caused all guilty of adultery to mount Ferrara's scaffold, since he had made his beloved Ugo die for that crime.[53]

Alone as this case stands in that Court-Almanac: History, as isolated is the like in that chronicle of humbler life: "The Newgate Calendar." Its single record shows that the mental state of the murderer, one Sheen, merited investigation. In addition to this feature, it illustrates the strange utility of strict construction of the "Letter of the Law" in providing a loop-hole for escape, where the most meritorious plea of Mental Science would, if it had not fastened the rope around the murderer's neck more quickly and surely, have been at best "laughed out of court." The parochial officers on learning who the father of the bantling was, induced him to marry its mother by a present of five pounds sterling. Sheen manifested an unaccountable antipathy to his infant. When it was presented to him, he turned away in disgust or in passion, repeatedly struck it and eventually he deliberately cut its throat. The child having been baptized William *Beadle* Sheen, its god-father and procurer of the marriage had unwittingly thus saved the murderer's life, for the indictment was drawn against the murderer of William Sheen (the father's name also) and therefore dismissed. On a second trial, the indictment having been correctly drawn, proceedings were definitely squashed on entry of the plea, *autre fois acquit*. The liberated murderer continued to reside at the scene of his crime, and repeatedly came before the magistrates upon allegations of riot and intoxication.

Among the agricultural classes, where this crime is sometimes committed by the father of a natural child, rare as it is, his part usually extends no further than abetting it. The instigator is in ninety-nine of a hundred cases the mother, and her accessory is often driven to do his part by threats of exposure or by some other imperative motive.

Whatever the foundation of a fatherly feeling, to which so few exceptions are found, Law can see in it only an inimical force, against whose encroachments constant vigilance and strict adherence to apparently harsh but wisdom-founded edicts are necessary. Often when the legislative chamber has been entered by law-makers who, as procreators of natural children were at the same time law-breakers, the barrier against law-breaking like their own has been undermined through special legislation, then breached by establishing vicious precedents, and finally borne down by a turbulent sea of crime and confusion, which thenceforth covered the domain of what had before been Law and Order.

So far am I from picturing thus on abstract assumption, that wishing to illustrate by what has been realized in the concrete I am at a loss—not for an example—but from the history of which nation, and from which several epochs in the history of the same nation, to select. One where the developing encroachment and consummation occurred with impressive rapidity, is that of Rome from the later mediaeval, down to the later part of the renascent period. The social and domestic life of the faction leaders—self-styled "nobili"—had become more thoroughly permeated with profligacy at this latter time than it had been — and sufficiently so — before. The laws of the *Rota Romana* became more and more relaxed in enforcement, and eventually the code in its parchments sanctioned the anticipating practice. One of the first codified departures was from the law, which treated the bastard, in the strict sense of the word, as *filius nullius*, and denied him succession to any and all estate on the paternal side. The new law while emphasizing the *nothus'* exclusion from inheritance of his father, allowed him to inherit from his brother: *Successio non degenator natis ex damnato coitu, quando non agitur de succedendo patri sed fratri.* I leave it to our legal friends to decide: how long would it take to capture the citadel of the paternal estate by the round about path of assault thus laid bare? The problem how one can be the brother, by common paternity, of the son of one who is not one's father, is not to be solved even with the assistance of the Seleucid and Ptolemaic family trees; although these present combinations, which fully realized, what is in the power of the most fertile fancy to imagine.

The chief changes were, however, made in local ordinances. The promulgators of these, gaining confidence from decade to decade of progressing demoralization, did not content themselves with merely relieving the disabilities of their favorites; they made privileged persons of them. The lives of legitimate citizens, not of noble birth (there were a few)were at the mercy of any nobleman's adventurous bastards, and for a reasonable consideration. An ordinance of 1580 decreed, that immunity for the crime of murder could be secured by a baron *or his bastard* (the latter is specifically designated!) on the payment of one thousand pounds (at 20 soldi each); by a knight or his "canaleretto" for four hundred. It so happened, that in the same year which witnessed the murder by a nobleman of his two daughters and his obtaining immunity at the price of eight hundred ducats, the bastard Franches-

etto Cibo, who practically governed Rome through his
father, secured the issue of a decree ordering the treasurer
to deliver to him all such fines as were in excess of one
hundred and fifty ducats — the blood moneys below that
limit, were permitted to go to the treasury. Thus was pre·
pared the soil on which the loathsome parasitical growth of
Borgias, Medicis, Roveres, Farneses and Riarios flourished;then
occurred excesses of the lower orders, recalling Sabine raid
and Marian tumult; and among the higher, orgies and crimes
which Elagabal and Commodus might have emulated. The
unloosening of all ties of fealty and honor, the substitution
of Borgia powders and bravo daggers for statesmanship
and warrior bravery, were local outcomes of this condition.
It went further; Macchiavelli's syllabus, inspired by bastard
models and favored by bastard patrons, carried the plague
across the Alps to poison the mind of the monarch who
ordered the fatal tocsin sounded on St. Bartholomew's day.
We can trace another transalpine-borne virus to Brinvil-
liers and her coterie, a gift of the Borgias. The perambu·
lating professor of their art, countryman of Macchiavel and
of a manufacturer of another· venom in pornographic
guise, Aretin, bore the suggestive name Exili. In these
three, the very Nadir of all that is repugnant to those of
clean hands, minds and morals seems to have been reached.
· No wonder is it, that bastardy in Italy not alone ceased to
be a disgrace, but under circumstances became a boast and
that even to the end of the last century, such sayings
were prevalent as: *Meglio un Gherardini bastardo che un
Corsini ben nato!*

The partiality shown natural children in condoning
faults and crimes, as well as in carrying generosity to them
to the limit of bestowing power, is commonly attributed to
the father's desire to make some restitution in repair of
the injury done an innocent offspring by the tainture for
which the father is responsible. A halo of romance is added
by sentimentalists, who affect to see therein a transfer or
continuance to the love-child of that affection, which its
parents had entertained for each other. Another common
view is fairly expressed by an eminent barrister, less known

however as a legal luminary than notorious as the butt of
his biographers, though at the same time distinguished as the
master biographer of another; Boswell in dissenting from
his eidolon Johnson's reprobation[54] of the "Letters"
of "Gentleman Chesterfield" to his (natural) son Stanhope,
says: "there was considerable merit in paying so much
attention to the improvement of one who was dependent
upon his lordship's protection; it has, probably, been
exceeded in no instance by the most exemplary parent, and
though I can by no means approve of confounding the distinction between lawful and illicit off-spring, which is in
effect insulting the civil establishment of our country, to look
no higher; I cannot help thinking it laudable to be kindly
attentive to those of whose existence we have, in any way
been the cause." Had the outpourings of generosity their source
in contrite desire to repair a wrong, they were in part direct-
ed thither, where restitution is equally if not more imperative,
and often much more urgent. But rarely are they allowed
to flow in that channel of equity. The neglect and — in
notorious instances — inventive contumely, meted out to
mothers of natural children, idolized by the seducers of
the former, tell a different tale; one showing the studied
favoring of such children to be due to naught but an ex-
aggeration of that feature intrinsic to paternal love, which dis-
tinguishes it from the maternal, namely: egoism, in vicarious
display. The mothers of natural children, on whom titles and
revenues were lavished or for whom special principalities
had even been established, remain unknown or are dismissed with the barest mention in history; which thus
reflects the treatment they experienced in life. There are
no Chesterfieldian "Letters by a Gentleman to the Mother
of his (natural) Son!" A Chancellor of France was
found obsequious enough to insert a clause in a fun-
damental ordinance of the realm, — or rather of Christendom
— which ensured oblivion[55] for the discarded mother of his
master's pampered bantlings.

Many have thought to find a strange inconsistency, at
best a foreign graft added to relieve an otherwise uncom-
promisingly abhorrent character, in the Moor of "Titus

Andronicus." It is where in a burst of fatherly affection, he rescues his *adulterinus* from contemplated slaughter, and later displays solicitude for, and pride in, the off-spring of his and Tamora's miscegenation. However, the great dramatist was as faithful to Nature here as elsewhere; parallel characters lived, and parallel transactions are recorded in the pages of actual history.[56] Caligula loved one object besides his horses: the *adulterine* Drusilla. Of this daughter who tortured, bit, scratched and tore or broke whatever of the animate or inanimate was in reach, Caligula seeing himself duplicated, thought, as Shakespeare makes the Moor express it:

"My mistress is my mistress; this myself,"
"The vigor and the picture of my youth,"
"This before all the world do I prefer;"
"This maugre all the world will I keep safe,"
"Or some of you shall smoke for it in Rome."

One of the best examples of a nobler kind of this infatuation is that of Franklin, already touched on. Even here the essential character of vicarious egoism is exhibited and happily epitomized by the biographer[57] who paints Benjamin as to William "at the same time his friend, his intimate and his easy companion." In other, less meritorious than notorious instances, paternal fondness for bastards appear, as a nauseating reflection and re-re-flection between the ego of the father and the reincarnation in an *alter ego*, which his vanity descries in the off-spring. It often becomes perverted into callous selfishness if it does not degenerate into almost insane vain glory. It is also combined with "sexual pride"; a factor of masculinity, more integral and influential than is realized when study of character remains exclusively a survey of its expanse and surface outgrowths, ignoring the intricate course and source of their roots. Had Alcibiades therein been less immodestly boastful, Leiotichides would have realized the paternal project and the Athenian race of Timea's seducer, mounted the Spartan throne of Timea's husband; but the father's indiscreet braggadocio furnished Agesilaus[58] the grounds needed to displace the *adulterinus*. Distinctly and dangerously near tipping were the scales of Marc Antony's mental balance, when, in defending the endorsement of the procreation of Cleopatra's twin bastards, he alleged the following: "A noble family can maintain itself solely by a large progeny of princely children. He, for example, was descended from Hercules, who had not risked staking his descent upon the offspring of any one female. For neither respecting the laws nor fearing the penalties of Solon concerning illegal impregnation, he had permitted nature's impulse to run its course, thus making him the founder of numerous septs." Consistently with this excursion into the realm of megalomania, Antony named the male of the twins, "Helios," and the female, "Selene." One effort more, in the same direction of nebular nomenclature, had designated the parentage of the celestial children and most appositely, in view of the bewildering maze and Tartarus obscurity of Ptolemaic genealogy, Chaos.

Under some circumstances this paternal pride is more reasonable and that is when bastards contrast so favorably with lawful offspring as they do in the records of regal and other prominent families. Placing side by side the names of their illegitimate representatives and of a near and cotemporary legitimate one, a glance at the resulting columns seems to prove an overwhelming superiority of the former:

BASTARD REPRESENTATIVE.	LEGITIMATE REPRESENTATIVE.
1. Arnulf, German-Roman Emperor; the first to defeat the Norsemen; rescued the Pope and carried Rome by assault; uniformly victorious over Slavonic and Scandinavian invaders as over rebellion and faction at home.	1. Charles the Fat, German-Roman Emperor *de jure* of the same (Carlovingian) family as Arnulf; sluggard, mentally incompetent and deposed for that reason.
2. Austria, Don Juan of Austria, destroyed Solyman's Armada which menaced European civilization, at Lepanto; well-nigh redeemed the Netherlands, though hampered by his brother's inventive stupidity.	2. Phillip II of Spain, his brother, sacrificed *his* Armada in the English channel; lost Spain, the Netherlands·
3. Berwick, Duke of Fitzjames; victor of Almanza, excellent diplomat and general; founded flourishing families both in Spain and in France.	3. Chevalier St. George, bigot, narrow-minded, unsuccessful claimant and leader; wrecked the rising of 1715; left the degenerate "Pretender" and the Cardinal York in whose persons the legitimate Stuart dynasty ended.

4. Buchan, Francis Stewart of; victor of Beaugé.

4. James King of Scotland, any one from the Third to the Seventh will do!

5. Charlemagne, Liberator of the Pontiff; conquered the Longbards and Saxons; united well-nigh the extent of the once Roman Empire under one sway.

5. Carlmann, his brother, monkish, easily led, and intimidated to abandon his rights of succession.

6. Robert Dudley, *adulterinus* by Lady Sheffield of the notorious Leicester; able, learned and left a distinguished—also bastard—son Charles.

6. "Lord Quondam," so nicknamed, an imbecile but legitimate successor to the estate from which Robert was excluded owing to impeached legitimacy.

7. Murray, Regent of Scotland, controlled its tumultuous people and factious nobles; through unprecedented difficulties interposed by religious conflict, foreign complications, his sister's and Cecil Burleigh's intrigues, he triumphed in the cabine and [at Langside] on the battle-field.

7. Mary Stuart, his sister, deposed, notwithstanding unusual factors and efforts in her favor, for crimes as shamelessly as stupidly committed; weak and a failure as a woman, she was not more successful as a would-be virago.

8. Douglass of Liddesdale, excellent military leader, successiul against the English.

8. Douglass the Fat and Douglass "Tineman;" these cognomina tell the tale.

9. Marshall Saxe; saved the land of his adoption at Fontenoy and Bergen-op-Zoom; he inherited the courtly graces, Herculean strength and creative imagination of his father, to which were added the military talents of the Königsmarks; notwithstanding his adherence—more from pride than enthusiasm—to a religious faith practically proscribed in France, he as a stranger gained the nigh royal rank of "Serene Highness," also a higher title to fame by writings which show him to have been a genius of the prophetic order.

9. August III Elector of Saxony and King of Poland, inherited all the faults, but neither the physical nor the mental gifts of his father; lost the influence of his house in Poland and his armies to Frederick the Great; his mother had been as bigoted a Lutheran, as his father was a renegade one; he paralleled his father in the latter respect as well as in political alliances—changing his religion adversely to his people and, equally *mal-a-propos*, deserting Frederick the Great when the latter was in the full tide of victory.

The voice of contemporaries echoed, the event justified, and the verdict of history confirmed the esteem in which his love child of chivalrous promise, Don Juan of Austria, was held by Charles the Fifth. That esteem could not but have become enhanced by his disappointment in legitimate Philip, whose being "out of harm's way" at St. Quentin's battle marred joy at its glorious victory. What more natural, than that deceived and betrayed by his wedlock-born weakling Henry, the pliant tool in foeman hands, Jerusalem's deliverer[59] should seek consolation in his fatherhood to knightly Manfred, faithful Enzio and Antioch-Conqueror Frederick? Aside from those of royal blood, bastards have been and are regarded by many, the superiors of legitimately born individuals; at least or notably in particular directions. Such view has found expression in popular sayings;[54] it has the support of countless confirmatory demonstrations in contrasts furnished by historical characters like those mentioned; finally it seems almost a presumption of Biology. The investigations whose fragmentary results I herewith offer,[60] were begun with anticipations determined by the latter, and became more and more strengthened by the second, until critical analysis caused the original bias to fade. The prevailing view does not appear invalidated regarding the surface features of the bastard nature and particularly its robust vigor. But as regards certain other and most important characters I found, that to attribute such superiority to the factors involved in illegitimate birth, were hasty; to do so exclusively, altogether erroneous. Usually the merits of bastards are, quantitatively considered—regarding the qualitative more anon—relative rather than absolute. Their comparison with kin of the full blood favors them to just that extent to which the former fail to attain the average standard. The thus increased discrepancy brings out such superior features as the nothus possesses in high relief. In the history of royal succession, the legally privileged competitor for fame enters the arena with all the heraldic pomp of the full blood, and the blazonry of legitimacy on the unstained shield, it is true; but he is otherwise so awkwardly trained and wretchedly equipped, that if the ordeal's outcome be at all surprising, it is so, because defeat is not a more uniform result, and a more crushing one. Language fails to adequately express the attitude of royal matchmakers. There have been instances which passive ignorance could not have produced, nor that colossal obstinacy which is pride- and precedent-grown, repeated, unless deliberate invention[61] in studied defiance of the simple laws of breeding, known to their humblest subjects, had been resorted to. If you can imagine a verdant tourist of the Stone Age suggesting the like to some shepherd breeder of primordial cattle, picture also the sequel, which, in default of a vocabulary sufficiently elaborate to express just indignation, would be that offended herdsman's supplementary remonstrance through the post-nuchal application of the handiest palaeolithic implement, to that foolhardy traveler. What the biologist in the laboratory and statistician in the library systemize, analyze and trace to finite causes, has been long, if crudely, known, through observations accumulated from the day of the hunter to that of the shepherd, and made in field and forest, by lake and stream. But those in high station thought themselves privileged to exert malaprop ingenuity in creating monstrosities, or as exempt from consequences which so many "terrible examples" should have warned them of. That these, in disregarding the warning, did not do so with impunity, is told in the sad record of royal degeneracy, whose lesson Ireland[62] summarizes by observing, that while the church had the power to grant dispensation from the consequences of violating Canonical laws it was beyond its power to grant such for violation of the laws of Nature.

The obstacles to attaining distinction were more than half overcome for Don Juan, the bastard son of Charles V, by his foil being such a creature as was Philip the Second. Berwick might for his brilliant career, contrasting with that of the "true eaglet"[63] that safely dares the sun," thank the Churchill blood in his veins. His wretched rival had not only not a drop of redeeming admixture to the blood of the paternal "hatchet face," but an increment of the very contrary, from the degenerate house of Modena. When to the sufficiently deteriorating influence of a monotonous series of marital unions limited to the affluent caste, was added the malign and degeneracy provoking one of consanguinity carried to the point of incestuous union, the result could not be otherwise than the procreation of the blase, the abulic, the imbecile or the cripple. On such a background almost any crossing of breed could not but produce a better offspring; worse had been out of realm of possibility, as the stage next beneath the one represented by such as Charles of progenic skull fame, Don Carlos, son of the second Philip, and Bavarian Otto is an alternative between a nonviable monstrosity; or Zero, by the merciful dispensation of barrenness. The story of competitors unfolded in such degeneracy telling cognomina as: the "Fat," the "Silly," the "Bald" the "Lazy" and the "Stammerer," makes the success of an Arnulf easy to understand.

Having opened the question of bastard character as to one of its aggressive manifestations, permit me to add the general result of inquiries on this head, before proceeding to their bearing on legislation. I found the available statistics barren of information or defective in necessary details. On the theory that a few cases well studied, promise more than thousands merely glanced at or forced into the Procrustean bed of an arbitrary system, I collected such from general and biographical history as well as from law records and reminiscences of cotemporaries. The fact soon forced itself on my attention that bastardy presents as many grades and as marked contrasts as wedlock procreation. The offspring of a *marriage de convenance* and of a union of affection differ; as much do those of certain classes of bastards differ from each other. I have therefore divided them into two groups, one containing three "benign" classes: A, composed of the children born in morganatic wedlock or so-called marriage by the left hand, with these I have included "mantle-children" and the offspring of surreptitious and stolen marriages; B, includes love-children, of which class, Erasmus, of Rotterdam, Pomponius, Manfred of Tarent and d'Alembert are types; C, comprises those born in open concubinage, *quasi* marriage "without the ring;" to this class belong, for example, Guy of Flanders, the Baryatinska, Duke of Grafton, Charles Martel, Lord Lovat's son Frazer, Smithson, Vertus, Douglas of Morton, John Corvinus, Margaret of Parma, Padella, Arsames, Mastanabal and Ferrand of Aragon   The other or Group II comprises classes in which the affection of the parents was more ignoble or less enduring, the mother frivolous, vulgar, meretricious, or, worst of all, added the guilt of marital infidelity to the common one of illicit procreation.

For reasons which will appear, I have in some tables discriminated between those natural children whose parents had been, one or both, themselves of illegitimate birth, or who procreated their like in turn, from those whose parents were of legitimate birth and those who are not known to have so procreated. Largely represented as are aggressive and serious crimes among the illegitimately born of the series, they are still more so among those who were bastards in a second generation, having had parents one or both themselves born out of wedlock. The figures are also higher amongst those who were the first in a series of two bastard generations and culminate in a group which, though a very small one, I have retained in separate place in some tables, since the culmination referred to, is parallel and consistent with the variation of other figures in the series, which rise and fall in harmony. It consists of those who, having had bastard born parents, represented double bastardy in their persons and, both receiving and giving, had transmitted the heirloom with interest to a third bastard generation. The individuals comprised in it, occupied a position in their family line from which they could, if Janus-headed, have enjoyed a symmetrical prospect and retrospect of bastardy like their own. Some of the worst monsters of history find their place here. The combined influence of bastardization involving greater venality, and of doubled as well as trebled tainture, is drastically illustrated in Table V; the ratio of murderers in the small category where all these sinister influences coincide is six times as great as that of the group where but one was operative, four times as great as the latter where one of them was absent, and so on. A glance at Table VII, based on a corresponding gauge of female morality shows results pointing in the same direction:

Immunity to the penalty of crime indicates on the part of an individual or a class of offenders: first, a cunning choice of the offense, the place and time at which, and persons against whom such offenses may be committed with impunity or with least chance of detection; second, a knowledge of so much of the law as may enable the crim-

inal to avoid its snares, and if snared, to take advantage of
loop-holes of escape, which indeed are often provided for, in
advance of, or in the very commission of the misdeed; third,
a faculty of dissimulating guilt; fourth, readiness, in des-
perate situations, to sacrifice confederates, if pardon or len-
ity be obtainable thereby—in short, it is not going far
wrong, to summarize the qualities useful in this direction as
rascality in its most visible form. Tables II, III and IV
compared, show, aside from the great immunity enjoyed by so
large a part of all offenders, that the group with the larger
ratio of offenders also has the larger proportion of immune
evil-doers. In addition to the preponderance of criminals in
Group II, it is found that among three hundred of simple
tainture, ninety-three persons had been guilty of either one or
several of the three felonies, murder, usurpation or sedition,
of whom thirty-one escaped all consequences of those or other
evil acts they may have committed; while of the division
of doubly taintured, ninety-eight in all, sixty-one offended
and twenty-nine of these escaped. For convenience I sub-
join these figures separately.

DIVISION T.		DIVISION TT.	
Committed Offences;.. $\frac{93}{300}$ or per mille	**310**	$\frac{61}{98}$ or per mille	**622**
Immunes $\frac{31}{93}$ or per mille Total of Offenders	**333**	$\frac{29}{61}$ or per mille	**475**
Immunes $\frac{31}{300}$ or per mille Total of Division	**103**	$\frac{29}{98}$ or per mille	**295**

The doubly taintured division has over twice as many
offenders and nearly three times as many of them immune
to consequences as the other. Of the [former, practically there escaped three for every two such in the division
of simple tainture (1,4:1,0.)

The number of these individuals removed from the
community through channels of a legal nature or in obedi-
ence to popular clamor and popular exigency is startling.
Out of 400 males 42 suffered death under the hands of public
executioners, or to evade such assistance committed suicide;
48 were victims of deliberate assassination or fell in riotous
uprisings against or rebellious conflict instigated by themselves. Of those who thus perished by violence, excluding

death on the battle-field, in duel, or through accident, the number is nearly evenly divided between those whose removal took place under the auspices of legally appointed functionaries and those in whose case the delaying formalities, incident to their co-operation had been dispensed with.[65] In addition many were banished, outlawed, fugitives from justice or imprisoned[see Table III]usually in mitigation through executive clemency of the extreme legal penalty of the law. The higher absolute figure of the eliminated, which suggests a greater degree of criminality in itself, but even still more so the, notwithstanding this greater proportion of immune offenders among the doubly taintured, indicate an intensification of those characteristics on which criminality rests; and as the product of multiplication represents multiplier and multiplicand, we can but conclude that as bastardy is a factor involved in both, that the latter either crops out of or takes up and carries with it a criminal disposition.[66] By this abused term I mean a condition of protoplasmic vigor, manifested in aggressive and assimilative restlessness, free from the restraints or most of those which are comprised in the inherited inhibitory mechanisms of the legitimately born, in civilized communities. As the multiplication of bastardy by bastardy is accompanied by a multiplication of criminality found in single bastardy, it is reasonable to assume that the two are closely intertwined and stand on a common foundation, or that, both growing in the same favoring soil, alike flourish with its expansion. It might be here interposed that the excess of criminals among those of double tainture need not necessarily be due to that as an intrinsic factor. As their overwhelming majority is found with Group II, which as a group necessarily be in itself contributes more evil-doers than Group I, the characters determining the greater degree of criminality being presumably shared by its sub-divisions, this preponderance, it could be argued, might be really due to the sinister influence of their adulterine, meretricious or clandestine procreation, and not to an intensification of predisposition thereto by the multiplied bastard tainture. This can be tested by separating each group, into divisions according to tainture. The resulting tables prove, as far as figures may, that the apparent rise of the criminal ratio in the aggregate of those doubly taintured is due to both factors, each acting independently though parallel with the other. Whatever the under-lying reasons are, whether they be identical or distinct, they are connected with the baser form of procreation in the one case, of multiplied bastardy in the other. When they coincide, an increment [Tables V and VI] is observed, in which both share, so that we may assume them to agree in their tendencies. The Table in which the crime of murder is used as guage, shows over twice as large a proportion of homicides among the doubly taintured as among the singly taintured of the benign form of bastardy. In the group of adulterine, meretrix-born and the offspring of clan-destine intercourse the doubly taintured division again shows over twice as many murderers as those of simple tainture. Now, if we contrast the murderers in Group II of simple tainture, with the corresponding division of the benign Group I, we find almost treble as many homicides. A lesser dis-crepancy exists in the other division, the one of the doubly taintured; but still more than a double one, nearly 23:10. The united influence of sinister birth of the malign kind and of double tainture elsewhere shown is exhibited in the ratio of the fourth rubric; it trebles the proportion of homicides. [See also Table V.]

Among thirty-seven females with double tainture, all but two of whom appertain to Group II, twenty-seven were of immoral character or had been involved in illicit relations at some time; twenty-six of these had procreated illegiti-mately. Of eighty-three with simple tainture, twenty-one had been likewise; four in the better Group I out of the thirty-one; and seventeen out of the fifty-two of the bad Group II. [Table VI]. These figures nearly correspond to the figures indicating male criminality.

What these Tables show as to the frequency of deeds of aggression by those of one sex, and such of indulgence by those of the other is confirmed by Tables II, III and IV relat-ing to criminality and immorality in general. Other features, such involving individualities and not susceptible of statistical demonstration, point in the same direction. But the strongest verification is found in the two endogenous ratio changes, which are explicable only if we assume that bastardy in itself,involves factors identical in origin with, or developing in the same way as those, which underlie criminality in the general population; but active in markedly higher degree. One of these ratio changes corresponds to the intensity of tainture, being higher with its darker shades, such as mark the off-spring of meretricious and adulterous intercourse, and lower with the lighter tainted, born of amatory and morganatic unions. The other rises and falls with the greater and lesser extent of the family line stained by illegitimacy, whether ascendant, descendant, or both. As the influence of their causes is parallel, their concurrence in the same subjects results in a cumulating rise to a point higher than the one reached when either operates singly. Thus far I have merely hinted that bastardy is one of the widest channels by which crime is disseminated and multiplied. It is thus active in other ways than the ones made apparent in these figures. Law therefore is as strongly justified in discouraging bastardy to the full extent of its power, as it is in eradicating the source of, or blocking the channels of access, by which any agency of evil threatens society.

Investigation and analysis of individual cases does not sustain the impression, so plausible *a priori*, that poverty, contemptuous treatment, sense of disgrace and exclusion from walks of life reserved for the legitimate are to any great extent accountable for the oblique courses so frequently followed by the sinister-barred. Relatively benign characters are found among the discarded and neglected, as often as demonic malignants develop from pampered children and patronized favorites.[67] The notorious monsters of bastard birth in history occurred almost exclusively among those who had been inviduously distinguished by fatherly fondness, or favored by fortune in other ways; a few like Elagabalus were spoiled pets of the populace. Opportunity offered by possession of wealth for multiplying mischief with the legitimate offspring of the affluent, has also been given to their illegitimate offspring, and with the like but more manifold results! The propriety of incorporating as one "Defective Class" the deformed in brain or body, the victims of disease or disaster, and those transgressing through misery or malice, doubtful as it seems to me on most grounds, appears particularly so in the light of this enquiry. In no way do its results sustain a relation between insanity and criminality, prominently as the latter crops out in them; in part they are antagonistic to that popularized dogma expounded by the *"uomo delinquente"* school.[68] The illegitimately born of healthy parents, are at least as robust as those legitimately born of like parents, nay rather more so. Opportunities for comparing the legitimate and illegitimate offspring of the same father are not few; and, where there exists a notable difference, psychically or physically, it is in the former respect more frequently in favor of the bastard, in the latter almost invariably so. The rare instances offering opportunites of comparing offspring procreated before, with such procreated after marriage, both parents being the same, confirm what is found in the former more numerous, though, as tests, less pure cases.

To be continued.

See Tables on following pages.

Footnotes will appear in January Number.

TABLE I.

Showing number of homicides among 398 males, illegitimately born.*

Classes	GROUP I.			Total of Classes A B & C	GROUP II.			Total of Classes D, E & F	Total of all Classes
	A. Morganatic.	B. Amatory.	C. Concubinage.		D. Clandestine.	E. Meretricious.	F. Adulterine.		
Total of illegitimately born males	35	60	101	196	68	84	50	202	398
Do. committed or attempted murder	2	4	15	21	17	30	16	63	84
Do. Do. percentage	$.05\frac{7}{10}$	$.06\frac{6}{10}$	$.14\frac{8}{10}$	$.10\frac{7}{10}$	.25	$.35\frac{7}{10}$	.32	$.31\frac{1}{10}$	$.21\frac{1}{10}$
Do. motives connected with dynastic or governmental questions and political ambition	2	3	14	19	14	12	4	30	49
Do. motives of a personal or vulgarly felonious nature	0	1	1	2	3	18	12	33	35
Ratio of two latter classes, per 100 total murderers	100:0	75:25	94:6	90:6	82:17	40:60	25:75	47:52	58:41
Murderers of their kin	2	2	7	11	9	5	7	21	32
Percentage: Murderers of kin among total of illegitimate	$.05\frac{7}{10}$	$.03\frac{3}{10}$	$.06\frac{9}{10}$	$.05\frac{5}{10}$	$.13\frac{2}{10}$	$.05\frac{9}{10}$	.14	$.10\frac{3}{10}$	.08
Do.: Ratio to total of Murderers	100:100	50:100	46:100	52:100	52:100	16:100	43:100	33:100	38:100

*The cases are classified according to the nature of the parental relations, as explained in the sequel.

TABLE II.

(Males only.)

	Group I.					Group II.					Ratio of Group II to Group I, the latter as a standard.	
	Murder.	Usurpation.	Rebellion.	Lesser or no offense.	TOTAL and Ratio pro mille.	Murder.	Usurpation.	Rebellion.	Lesser or no offense.	TOTAL and Ratio pro mille.	I.	II.
Chief crime if any committed by sufferer												
Executed	3	0	9	3	76 / 15:1000	15	0	8	1	118 / 24:1000	1.00	1.55
Assassinated	3	2	0	7	61 / 12:1000	21	0	4	11	174 / 36:1000	1.00	2.85
Suicides	1	0	0	2	15 / 3:1000	1	1	3	2	34 / 7:1000	1.00	2.26
Total of those whose career terminated by a violent death, excluding those dying on battle fields, in duello, or by accident.	7	2	9	12	153 / 30:1000	37	1	15	14	331 / 67:1000	1.00	2.16

TABLE III.

(Males only.)

Chief offense committed	GROUP I					GROUP II					
	Murder.	Usurpation.	Rebellion.	Lesser or no offense.	TOTAL and RATIO.	Murder.	Usurpation.	Rebellion.	Lesser or no offense.	TOTAL and RATIO.	
1. Imprisoned, banished or fugitives from their native land..........	1	1	8	5	15: $\frac{76}{1000}$	8	1	5	8	22: $\frac{108}{1000}$	1,00 : 1,42
2. No vicissitudes nor penalties known to have been undergone........	12	12	3	124	151: $\frac{770}{1000}$	19	9	5	80	113: $\frac{559}{1000}$	1,00 : 0,72
3. Total of those in this and Table II.....	20	15	20	141	196*	64	11	25	102	202	
4. Ratios of third rubric.........	$\frac{102}{1000}$	$\frac{76}{1000}$	$\frac{102}{1000}$			$\frac{316}{1000}$	$\frac{54}{1000}$	$\frac{123}{1000}$			
5. Total of eliminations including those removed by violent means (Table II) with those in first rubric of this table.......	8	3	17	17	45: $\frac{229}{1000}$	45	2	20	22	89: $\frac{440}{1000}$	1,00 : 1,92

*The lesser offenses were not computed at this time, so they appear in Table IV, where the preponderance of Group II as to low felonies is one of the factors manifest.

TABLE IV.

Showing degree of relative immunity to consequences of felonies, among the offenders in 518 illegitimately born males and females.	Aggregate of Classes A, B, C, D.	Do., of Adulterine and Meretrix-born (E and F)	RATIO.
1. Committed offenses while *in a situation** to be held amenable to existing laws*	46	61	
			1,00 : 2,11
2. Ratio	144/1000	305/1000	
3. Suffered adequate penalties	27	16	
			1,00 : 0,44
4. Ratio	586/1000	262/1000	
5. Evaded such by flight, turning States Evidence, shifting guilt to others or incurred ridiculously inadequate punishments	10	16	
6. Escaped all and any consequences of their crimes	9	33	
			1,00 : 2,76
7. Ratio of latter to total of offenders....	195/1000	540/1000	
			1,00 : 5,89
8. Do., to total of classes	28/1000	165/1000	
			1,00 : 1,94
9. Ratio of aggregate, entirely or partly immune to total of offenders	413/1000	803/1000	
			1,00 : 4,10
10. Do., to total of classes	59/1000	242/1000	

*Obviously, it were faulty to include those whose immunity was due to monarchial or governmental position. Were these included the figures of Group II would not be so markedly in disfavoring contrast, for crimes committed from political motives, like the summary execution at the order of Ptolemy nothus (Auletes) of Berenice, the starvation of Rothsay by Albany and secret execution of Crispus, bear a larger proportion in Group I. In Group II, homicide, for example, appears more frequently the expression of vulgar, felonious motives.

E. C. Spitzka.

TABLE V.

Adulterine (F) and Meretrix-born (E) aggregate of males

	Whose parents were one or both of them illegitimately born,			Whose parents were reputed of legitimate birth,			TOTAL	Aggregate of remaining Classes A, B, C and D, viz: 'Morganatic,' 'Amatory,' 'Concubinage' and 'Clandestine.'
	procrea'd had not a third generat'n of bastards.	so procrea'd	Total	procrea'd a second generat'n of bastards.	not known to have so procrea'd	Total		
Total in Category	6	17	23	21	78	99	122	276
Homicides, simple	2	4	6	3	15	18	24	18
Homicides guilty of usurpation	3	2	5	2	1	3	8	14
Homicides guilty of treason, rebellion or sedition	4	4	·	4	6	10	14	6
Total of Homicides	5	10	15	9	22	31	46	38
Proportion *pro mille*	833	588		427	296		137	
Ratio crudely expressed, proportion in last column as the unit	6	4		3	2		1	

TABLE VI.

Showing Influence of Extent of Tainture on Sexual Morality, in 120 Illegitimately Born Females.

	GROUP I. Including Classes, "Moralfanatic," "Amatory" and "Concubinage."			GROUP II. Including Classes, "Clandestine," "Meretricious" and "Adulterine."			Total of Female Illegitimates.			Ratio of Groups; II as the unit in upper line; I as the unit in lower line.
	Total.	Unchaste.	Per Cent.	Total.	Unchaste.	Per Cent.	Total.	Unchaste.	Per Cent.	
Division T "Simple Tainture." a Total in group......	31			52			83			39 : 100
b Of Unchaste Character		4			17			21		100 : 252
c Percentage of Latter			$.12\frac{9}{10}$			$.32\frac{6}{10}$			$.25\frac{3}{10}$	
Division T T "Double Tainture." a As Above......	2			35			37			vacat
b "	0				27			27		0 : 100
c "						$.77\frac{1}{10}$			$.72\frac{9}{10}$	
Total of Females. a "	33			87			120			23 : 100
b "		4			44			48		100 : 417
c "			$.12\frac{1}{10}$			$.50\frac{5}{10}$			.40	
Procreated in Adultery or had Adulterous Relations	1		.25	16		$.36\frac{3}{10}$	17		$.35\frac{4}{10}$	
Procreated or had Liaisons without Breach of Wedlock.	3		.75	28		$.63\frac{6}{10}$	31		$.64\frac{5}{10}$	

Table VII.

	GROUP I. Morganatic, Amatory & Concubinage.			GROUP II. Clandestine, Meretricious and Adulterine.			TOTAL MALES.			Ratio of Homicides of Group I to do. Group II.
	Total.	Homicides.	Percen'ge.	Total.	Homicides.	Percen'ge.	Total.	Homicides.	Percen'ge.	
T Division of single-taintured	169			131			300			100 : 294
Do. Homicides		15			34			49		
Percentage of Homicides			$.08\frac{8}{10}$			$.25\frac{9}{10}$			$.16\frac{3}{10}$	
TT Division of Doubly-taintured	27			71			98			100 : 228
Do. Homicides		5			30			35		
Percentage of Homicides			$.18\frac{5}{10}$			$.42\frac{2}{10}$			$.35\frac{7}{10}$	
Total Males	196			202			398			100 : 309
Do. Homicides		20			64			84		
Percentage of Homicides			$.10\frac{2}{10}$			$.31\frac{6}{10}$			$.21\frac{3}{10}$	
Ratio of Homicides per centum in Division of single-taintured to that of double-taintured, the former assumed as standard.	100:210			100:162			100:212			